A
LET'S-READ-AND-FIND-OUT S

follow your nose

Smells give us all kinds of information. They can warn of danger; but they may also tell us when breakfast is ready or when a barbecue is being cooked.

In this gay book Mr. Showers carefully gives young readers an understanding of the mechanics of smelling and other functions of the nose, such as filtering and preheating of air that enters the lungs.

Paul Galdone's practical knowledge of children and his sense of humor enable him to illustrate the story with zest and enthusiasm. He catches reactions to gentle sniffs, to pleasant odors, and of course he shows some children with "noses in the air." Althogether a delightful book full of interesting and worthwhile science information.

follow

your nose

by PAUL SHOWERS
illustrated by Paul Galdone

THOMAS Y. CROWELL COMPANY New York

This Crowell Crocodile is one of the quality paperback editions
selected from Crowell's highly recommended:

―――― LET'S-READ-AND-FIND-OUT SCIENCE BOOKS ――――

Editors: Dr. Roma Gans, Professor Emeritus of Childhood Education, Teachers College, Columbia University
Dr. Franklyn M. Branley, Astronomer Emeritus and former Chairman of the American Museum—Hayden Planetarium

A Baby Starts to Grow
Bees and Beelines
Birds Eat and Eat and Eat
A Drop of Blood
Follow Your Nose
Hear Your Heart
High Sounds, Low Sounds
How a Seed Grows

How You Talk
It's Nesting Time
Ladybug, Ladybug, Fly Away Home
My Five Senses
My Visit to the Dinosaurs
Oxygen Keeps You Alive
Straight Hair, Curly Hair
The Sunlit Sea

A Tree Is a Plant
Use Your Brain
Water for Dinosaurs and You
What I Like About Toads
What Makes Day and Night
What the Moon Is Like
Why Frogs Are Wet
Your Skin and Mine

Copyright © 1963 by Paul Showers. Illustrations copyright © 1963 by Paul Galdone.
All rights reserved. Published in Canada by Fitzhenry & Whiteside Limited, Toronto.
Manufactured in the United States of America.
L.C. Card 63-15097. ISBN 0-690-00635-7.
2 3 4 5 6 7 8 9 10
CROWELL CROCODILE EDITION, 1975

What can your nose tell you?
My nose tells me a lot of things.

In the morning my nose tells me about breakfast.

I can smell toast.
I can smell my father's coffee.
I can smell the bacon my mother is broiling.

I dress in a hurry and

Outside, my nose tells me
 there are roses next door.
I cannot see the roses
 but I can smell them.

I know Mr. Carter
is painting his porch.
I can smell the paint.

I know when someone is having a cookout
  on our street.
I smell the smoke and I
  follow my nose to the cookout.

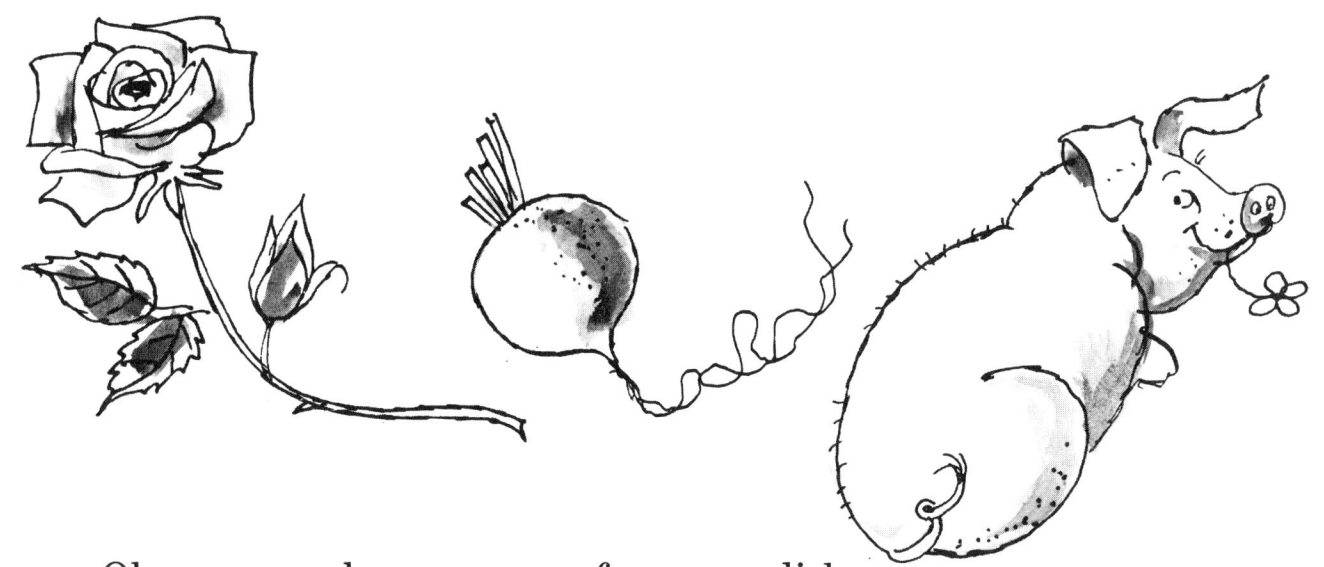

Oh, my nose knows a rose from a radish,
My nose knows a pig from a pie.
And my nose can tell
Just by the smell
When a skunk is passing by.

You smell through your nose.
Air goes into your nose through your nostrils.
Nostrils are the two little openings
   at the end of your nose.
The air goes up your nose and into your head.
When it gets into your head, you can smell it.
From your head the air goes down to your lungs.

Take a big breath through your nose.
The air goes into your head,
and down into your lungs.
Your lungs fill up with air.

Now let the air out. Keep your mouth closed.
Breathe out through your nose.

You do not smell the air when you breathe out.
You smell it only when you breathe in.

You can breathe in air through your mouth.
But can you smell when you do that?
Let's try it and see.

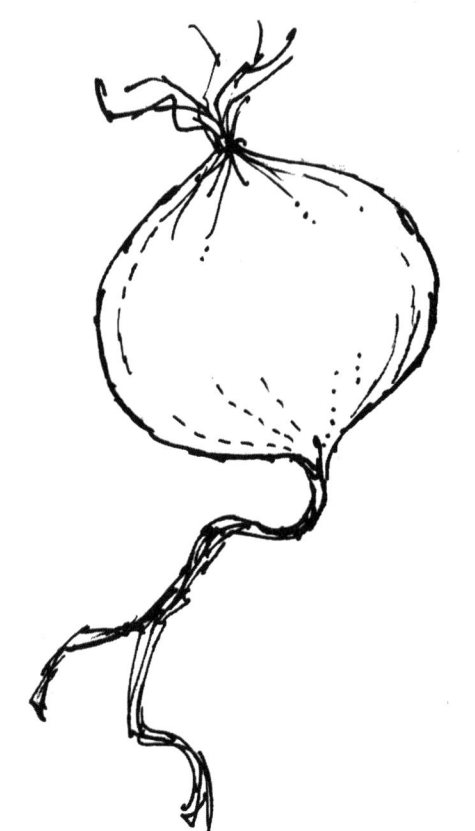  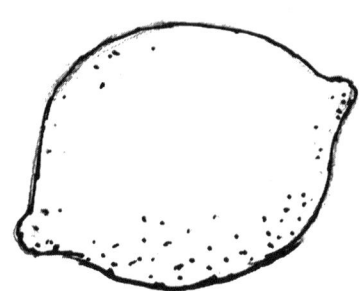

Take something that has a smell.
Take an onion—or a bar of soap—or a lemon.

Close your mouth.
Hold the lemon in front of your nose.
Breathe in through your nose.
Can you smell the lemon?

Now hold your nose so that you cannot breathe through your nostrils.
Hold the lemon in front of your mouth.
Breathe in through your mouth.
Can you smell the lemon?

You cannot smell when you hold your nose.
Can you taste when you hold your nose?
Let's find out.

Put four clean cups on the table.

Pour grape juice in the first cup,
tomato juice in the second cup,
orange soda in the third cup,
lemon soda in the fourth cup.

Take a sip of tomato juice and a sip of grape juice.
They taste different.

Take a sip of orange soda and of lemon soda.
It is easy to tell them apart.

Now hold your nose.
Breathe through your mouth.
You can't smell anything.

Keep holding your nose and sip the orange soda.
    Sip the lemon soda.
    Then the grape juice.
    Then the tomato juice.

What happens when you hold your nose?
What happens when you can't smell?

Things lose much of their taste.
Lemon soda and orange soda are not very different
 when you hold your nose.
Tomato juice and grape juice lose
 much of their taste when you hold your nose.

Your nose does other things.
It cleans the air that you breathe.
Inside your nose there are many hairs.
They are small, like the hairs on your arm.
They are close together.

When you breathe in, the air moves through the hairs.
The hairs catch the dust that is in the air.

They keep the dust out of your lungs
as a screen keeps flies out of the house.

Your nose also helps to warm the air you breathe.
If you breathe through your mouth,
   the air goes right into your lungs.
On a cold day, the cold air makes you feel cold inside.

When you breathe <u>through your nostrils</u>,
   the air goes up in your nose.
The air is warmed as it goes down into your lungs.

Your nose helps you in another way.
It helps you talk.
You cannot make the sound *M* without using your nose.
You cannot make the sound *N* without using your nose.
Try it and see.

Hold your nose.
Try to say *M*.
Try to say *N*.
The letters don't sound right.

When you have a cold, air cannot get through your nose.
Then you can't smell or taste.

I hate to have a cold,
    It isn't fun to eat.
I cannot smell my muffins,
    I cannot taste the meat.

There's just one thing about a cold
    That's really good, I think.
I cannot taste the medicine
    My mother makes me drink.

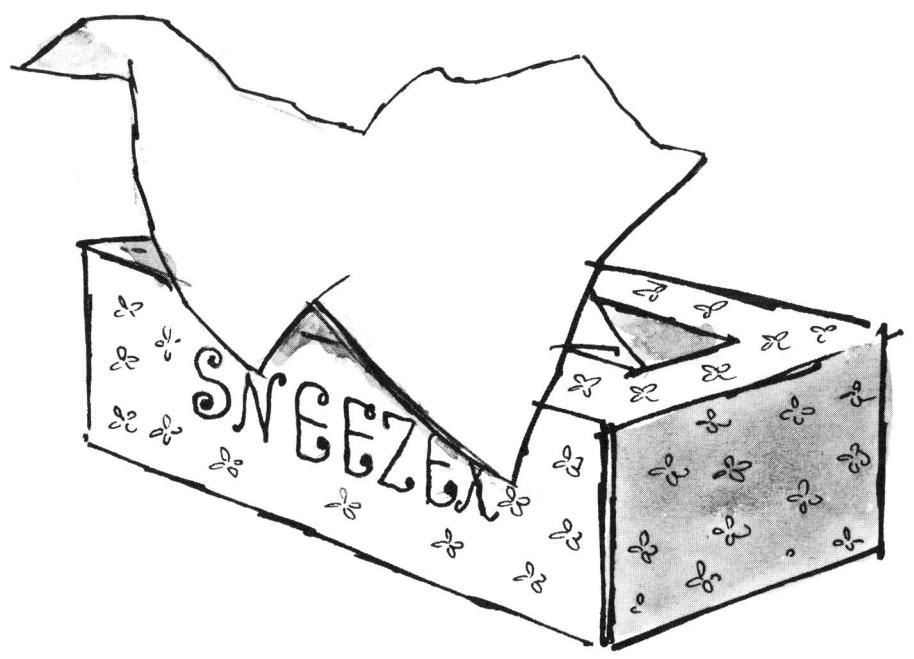

When you have a cold in your nose,
   you can't talk very well.
You can't say these verses very well.
This is how they sound when you have a very bad
   cold:

> I hate to have a cold,
>   It izid fud to eat.
> I caddot sbell by buffids,
>   I caddot taste the beat.
>
> There's just wud thig about a cold
>   That's really good, I thigk.
> I caddot taste the bedicid
>   By buther bakes be drigk.

ABOUT THE AUTHOR

Paul Showers is a newspaperman and writer. Starting out as a copy editor on the *Detroit Free Press* and later the *New York Herald Tribune*, he spent the war years as a sergeant on the staff of *Yank*, the Army weekly. Subsequently he worked briefly for the *New York Sunday Mirror*. He is now on the staff of *The New York Times Magazine*.

Mr. Showers was born in Sunnyside, Washington, and grew up in various parts of the country, among them a suburb of Chicago, Muskegon and Grand Rapids in Michigan, and Rochester, New York.

ABOUT THE ILLUSTRATOR

Paul Galdone is considered one of the outstanding illustrators of children's books. He studied at the Art Students League of New York and with George Grosz, and he spent his spare time sketching from life in parks, zoos, subways, and streets, and doing New England summer landscapes on his vacations.

Mr. Galdone is a native of Budapest, Hungary. He now lives in Rockland County, New York, with his wife and two children.